COLLABORATIVE ACTIVITIES MANUAL

PATRICK DEFAZIO
Onondaga Community College

WITH ORIGINAL MATERIAL BY

IRENE DOO
Austin Community College

INTERMEDIATE ALGEBRA

TENTH EDITION

Marvin L. Bittinger
Indiana University Purdue University Indianapolis

Boston San Francisco New York
London Toronto Sydney Tokyo Singapore Madrid
Mexico City Munich Paris Cape Town Hong Kong Montreal

Reproduced by Pearson Addison-Wesley from electronic files supplied by the authors.

Publishing as Pearson Addison-Wesley, 75 Arlington Street, Boston, MA 02116.

 Printed in the United States of America.

ISBN 0-321-30582-5

1 2 3 4 5 6 OPM 09 08 07 06

CONTENTS

Name Section Date

Activity R.3 Use the order of operations to modify an expression.

Focus	Order of operations
Time	10–15 minutes
Group size	2
Background	Understanding the rules for order of operations is essential to your study of algebra. This activity provides you with a different perspective on using the rules, and should help you learn the order of operations.

1. Write down the rules for order of operations in the space below. Refer to Section R.3 in your textbook, if needed.

2. The object of this activity is to insert grouping and/or operation symbols within a display of numbers in order to obtain a predetermined result. The symbols allowed are: ( ), +, –, ·, and ÷. For example, let's start with the display shown below.

$$1 \quad 2 \quad 3 \quad 4 \quad 5$$

When you insert the symbols as follows,

$$(1 + 2) \div 3 + 4 \cdot 5,$$

the result will be 21. Check that this is correct by simplifying the expression.

Now, practice on the following problem:

Where would you insert symbols to get a result of 16?

$$1 \quad 2 \quad 3 \quad 4 \quad 5$$

Check your answer by simplifying the expression.

3. Now, prepare an exercise for your partner to solve. Select five single-digit numbers (from 1 to 9) for display. Then, privately, insert grouping symbols and/or operations within your display, and calculate the result. Write the result and the display of numbers on a blank sheet of paper, and exchange papers with your partner.

   On the paper you receive, insert symbols so as to make the display equal the number given. When you are done, exchange papers again. Check your partner's solution with yours. Discuss any differences you find.

4. If you have time, prepare another exercise for your partner to solve.

Conclusion	The process of inserting symbols into a display of numbers should give you a good understanding of the rules for order of operations. It is also fun to manipulate symbols to achieve a predetermined goal.

Name Section Date

Activity R.6 Simplify algebraic expressions as a group.

Focus	Simplifying algebraic expressions
Time	10–20 minutes
Group size	3
Background	This activity will give you practice simplifying expressions as a group. There is usually more than one correct sequence of steps for simplifying an expression. Thus, it is important to be able to follow someone else's steps even if his or her approach is not like yours.

1. Each group member should select a different one of the following expressions.

$$4\{[8(x-3)+9]-[4(3x-7)+2]\}$$

$$3\{[6(x-4)+5^2]-2[5(x+8)-10^2]\}$$

$$7b-\{5[4(3b-8)-(9b+10)]+14\}$$

   Write the expression on a blank sheet of paper, and perform the first step of the simplification.

2. When you are done, pass your paper to the group member on your left. Look at the paper you receive and check the work shown. If you spot an error, discuss it with the group member who did the step. He or she should then correct the error.

3. Perform a second step of the simplification for the expression you received. Then pass the paper to the group member on your left.

4. Check the work on the problem you receive. Discuss and correct any errors as before, then do the next step. Continue passing the problems until all expressions have been simplified.

5. As a group, look at the final expression for each problem. When you are satisfied with the answers, compare them with the other groups in the class.

Conclusion	Simplifying expressions as a group gives you a better understanding of the simplification process. You can also use this process to solve linear equations, or to solve other types of equations.

Name Section Date

Activity 1.1 Identify linear equations as having one solution, no solutions, or infinitely many solutions.

Focus	Solving and identifying types of linear equations
Time	20 - 25 minutes
Group size	3
Background	This activity will help to strengthen your understanding of equations and reinforce your solving skills through identifying how many solutions a linear equation has.

1. Each group member should choose one of the three linear equation types listed below.

   Linear Equations with One Solution

   Linear Equations with No Solutions

   Linear Equations with Infinitely Many Solutions

   Write your selected equation type at the top of a sheet of notebook paper.

2. Each group member should now select a pair of expressions from the list below (choose one from Group A and one from Group B). Once you have chosen two expressions create an equation by writing them on the same line of your notebook paper with an equals sign "=" between them.

Group A	Group B	
$x-3+x+5$	$\frac{1}{2}(2x+12)+2x$	$2(x-3)+(x-11)$
$3x-(3+2x)$	$\frac{1}{3}(3x+15)-4$	$2(2x-(x+2))$
$3x+6$	$2x-1-x-2$	$2(x+1)$

3. Use the methods in Section 1.1 to solve the equation on your paper.

4. Determine if your equation is one of the type you selected in step 1. If it is, write your equation in the appropriate area of Results Table (below) and wait for your group members to fill in their areas. If it is not, create a new equation on your notebook paper by selecting a different pair of expressions from the list in step 1 and repeat steps 3 and 4. You should continue to select expressions until you discover the type of equation that you selected in step 1.

5. When all group members have entered their equations in the Results Table, each should select a different type of linear equation from the list in step 1. Complete steps 2 through 4 for this type of equation. Be sure to find a different equation to enter in the Results Table than the one already there.

6. Repeat the process one more time for the remaining type of linear equation.

Results Table

Types of Linear Equations	Equations You Have Found
Linear Equations with One Solution	
Linear Equations with No Solutions	
Linear Equations with Infinitely Many Solutions	

Conclusion	Understanding linear equations is very important in mathematics. Learning how to recognize the number of solutions a given linear equation has will strengthen your confidence when solving linear equations and enhance your understanding of what an equation is.

Name Section Date

Activity 1.1 Solve linear equations as a group.

Focus	Solving linear equations
Time	10–20 minutes
Group size	3
Background	This activity will give you practice with solving equations as a group. There is usually more than one correct sequence of steps for solving an equation. Thus, it is important to be able to follow someone else's steps even if his or her approach is not like yours.

1. Each group member should select a different one of the following equations.

$$9(4x+7)-3(5x-8)=6\left(\frac{2}{3}-x\right)-5\left(\frac{3}{5}+2x\right)$$

$$6\left[4(8-y)-5(9+3y)\right]-21=-7\left[3(7+4y)-4\right]$$

$$\frac{2x-5}{6}+\frac{4-7x}{8}=\frac{10+6x}{3}$$

Write the equation on a blank sheet of paper, and perform a first step of the solution.

2. When you are done, pass your paper to the group member on your left. Look at the paper you receive and check the work shown. If you spot an error, discuss it with the group member who did the step. He or she should then correct the error.

3. Perform a second step of the solution for the equation you received. Then pass the paper to the group member on your left.

4. Check the work on the problem you receive. Discuss and correct any errors as before, then do the next step. Continue passing the problems until all equations have been solved.

5. As a group, look at the solutions to all three equations. When you are satisfied with the answers, compare them with the other groups in the class.

Conclusion	Solving equations as a group gives you a better understanding of the solution process. You can also use this technique to simplify algebraic expressions (see Activity R.6) or to solve linear inequalities later on in this chapter.

Name Section Date

Activity 1.3 Use the five-step problem solving strategy.

Focus	Problem solving
Time	15 minutes
Group size	5
Background	The five-step strategy given in Section 1.3 of your textbook is used to help solve various types of applied problems. This activity gives you practice using this strategy to solve a problem different from those in the text.

Statement of the problem:

> Two members of your group are celebrating their birthdays today. The entire group goes out to lunch, and each birthday celebrant gets treated to his or her lunch by the other *four* members. Thus, if group members A and B are both celebrating their birthday, then A gets a free lunch but pays for part of B's lunch, and B gets a free lunch but pays for part of A's lunch. The other three group members pay for their own lunches as well as part of A's and B's lunches. Each meal costs the same amount and the total bill is $40. Ignore the amount for the tip.

1. Determine, as a group, how much each group member should contribute towards the bill. In the space below, explain how your group arrived at the solution. Use complete sentences and/or equations, and include enough detail so another group can understand your steps. Refer to the five steps for problem solving given in Section 1.3 of your textbook for assistance.

2. Exchange papers with another group. Analyze the strategy used by the other group. Did they use the five steps to solve the problem? If your group does not follow any of their steps, discuss this with the other group.

   Return the paper to the other group before continuing to the next step.

3. Now, suppose the total bill was \$60, and each meal costs the same. The rest of the problem is the same as before. Use your group's strategy from step 1 to solve this problem. Write the solution in the space below.

4. Finally, generalize the problem, and let each group member's meal cost x dollars, with the total bill of $5x$. Solve this problem as a group by using the strategy from step 1. Write your steps below. When you are done, exchange papers with another group, and compare their solution with your group's solution.

Conclusion	The five-step strategy for solving problems can be used to solve many kinds of problems, even one as unorthodox as the birthday lunch problem. When you encounter such problems in your daily life, don't forget that you can use the five steps to help you solve the problem.

Name Section Date

Activity 1.5 Solve inequalities and compound inequalities and present the solutions in multiple forms.

Focus	Compound inequalities, interval notation, set notation, graphs of intervals
Time	15 – 20 minutes
Group size	4
Background	The solutions to inequalities and compound inequalities can be represented in different ways. This activity will give you some practice solving inequalities and compound inequalities and converting solutions from one form to another.

1. Each group member should choose one of the inequalities listed below. Solve the inequality on a sheet of notebook paper. Write your answer in set notation (for example, $\{\, x \mid x \le -2 \,\}$).

$$0.1(60x-40) \ge 2$$

$$4(2y-6) \le 2y+6$$

$$\frac{1}{2}(2y+4)-3 > 9$$

$$5(2-x) < -3x+1$$

2. Pass your sheet of notebook paper to the group member on your left. Take the sheet that is handed to you and review the work. If the work appears to be correct, then graph the solution. If the work appears to be incorrect, work with the group member to correct it before graphing the solution.

3. Pass the sheet to the group member on your left. Take the sheet that is handed to you and review the work. If the work appears to be correct, then write the solution in interval notation (for example, $(-\infty,-2]$). If the work appears to be incorrect, work with the group member to correct it before writing the solution in interval notation.

4. Pass the sheet to the group member on your left. Take the sheet that is handed to you and review the work. If the work appears to be correct, then describe the solution in words (for example, "all real numbers less than and including –2"). If the work appears to be incorrect, work with the group member to correct it before describing the solution in words.

5. Now we will move on to complex inequalities. Each group member should choose one of the inequalities listed below. Solve the inequality on a new sheet of notebook paper. Write your answer in set notation (for example, $\{ x \mid x \leq -2 \text{ or } x > 3 \}$).

$$-10 \leq 2x + 4 < 5$$

$$-1 < \frac{6x-3}{2} \leq 2$$

$$2x + 7 < 17 \ \text{or} \ \frac{1}{2}(2x+6) \geq 10$$

$$10 - 3x > 15 \ \text{or} \ 0.2x + 1.1 \geq 3.1$$

6. Pass your sheet of notebook paper to the group member on your left. Take the sheet that is handed to you and review the work. If the work appears to be correct, then graph the solution. If the work appears to be incorrect, work with the group member to correct it before graphing the solution.

7. Pass the sheet to the group member on your left. Take the sheet that is handed to you and review the work. If the work appears to be correct, then write the solution in interval notation (for example, $(-\infty, -2] \cup (3, \infty)$). If the work appears to be incorrect, work with the group member to correct it before writing the solution in interval notation.

8. Pass the sheet to the group member on your left. Take the sheet that is handed to you and review the work. If the work appears to be correct, then describe the solution in words (for example, "all real numbers that are less than and including –2 or greater than 3"). If the work appears to be incorrect, work with the group member to correct it before describing the solution in words.

Conclusion	Being able to convert solutions of inequalities from one form to another strengthens your understanding of what these solutions are.

Name Section Date

Activity 1.6 Create and solve equations with absolute value as a group.

Focus	Solving equations with absolute value
Time	15–20 minutes
Group size	5
Background	Equations with absolute value are solved using the absolute-value principle, as shown in Section 1.6 of your textbook. This activity gives you practice in using this principle to solve equations with absolute value.

1. Each group will create and solve equations with absolute value by following the steps outlined below. Study the example so you understand the mechanics of this activity.

		Example
Step 1	The first group member thinks of a number, writes x = the number on a piece of notebook paper, and passes the paper to the second group member.	-2; $x = -2$
Step 2	The second group member multiplies both sides of the equation by a number. Write the new equation below the first one, and pass the paper to the third group member.	$3x = -6$
Step 3	The third group member adds or subtracts a number to both sides of the equation, and writes the new equation below the second one.	$3x - 2 = -8$
Step 4	The fourth group member takes the absolute value of both sides of the equation, and writes the resulting equation below the third one. Fold over the paper so that only the last equation is shown, then pass the paper to the fifth group member.	$\lvert 3x - 2 \rvert = 8$
Step 5	The fifth group member writes the equation on his or her record sheet, and solves the equation. There should be two solutions to the equation. Unfold the notebook paper, and check that one of the solutions matches the equation written by the first group member. If it does not, the group should find out where the mistake occurred.	$3x - 2 = 8$ *or* $3x - 2 = -8$ $3x = 10$ *or* $3x = -6$ $x = \frac{10}{3}$ *or* $x = -2$

2. Now you are ready to create your own equations. Each group member begins with step 1, writing down the equation on a piece of notebook paper. Pass the paper to the group member on your right.

   Modify the equation you receive according to step 2, then pass the paper to your right again. Continue until you reach step 5; you should have an equation that looks like $|ax+b|=c$. Write down this equation in the space below. Then, solve and check the equation, following the directions for step 5. When all group members are done, examine all five equations in your group and correct any errors.

3. Now, examine the two solutions in the equation that you solved. One of these solutions is the original equation written in the first step. Take the other solution, and do the same modifications that were done to the original equation. Compare the resulting equation with the one you got earlier. Are they identical? Why do you think this is so? Use complete sentences in your answer.

4. Compare your results with those from your fellow group members. What conclusions can you make about the two solutions to each absolute value equation?

Conclusion	Notice that you are writing equivalent equations in steps 2 to 4. This is one of the fundamental concepts in equation solving, as explained in Section 1.1 of your textbook. This activity should also give you a clearer understanding of how the absolute-value principle works, and consequently increase your proficiency in solving equations with absolute value.

Name Section Date

Activity 2.5 Practice graphing and identifying the graphs of linear equations.

Focus	Graphs of linear equations
Time	15–20 minutes
Group size	4
Background	In Section 2.5 of your textbook, linear equations were graphed using the slope and y-intercept. In Exercise Set 2.4 of your textbook, you were asked to find the slope of a line when given the graph of the line. This activity gives you practice graphing linear equations quickly using the slope and y-intercept, as well as practice deducing the slope-intercept equation of a line from its graph.

1. On each of four scraps of paper, write one of the following slopes.

   $m = \frac{2}{3}$ $\qquad m = \frac{3}{4}$ $\qquad m = -\frac{2}{3}$ $\qquad m = -\frac{3}{4}$

   Place the scraps of paper face down on the desk, and mix them up. Then, one by one, each group member should select one of the pieces of paper. Do not show your slope to the other members of your group.

2. Similarly, on each of four different scraps of paper, write one of the following y-intercepts.

   $b = -3$ $\qquad b = -1$ $\qquad b = 1$ $\qquad b = 3$

   As you did above, place the scraps of paper face down on the desk, and mix them up. Then, one by one, each group member should select one of the pieces of paper. Do not show your y-intercept to the other members of your group.

3. Write your linear equation formed by the values of m and b selected in the space below. Use the slope-intercept form $y = mx + b$. Do not show this equation to your fellow group members.

4. Graph your equation on the grid on the next page. Use the slope and y-intercept as shown in Section 2.5 of your textbook. Make sure the graph is neat, but do not label the graph with the equation.

5. When all group members have finished, place the graphs in a pile. The group should then agree on the correct equation for each graph *with no help from the person who drew the graph.* Find the slope and y-intercept of each graph, and use these values to form the proposed equation.

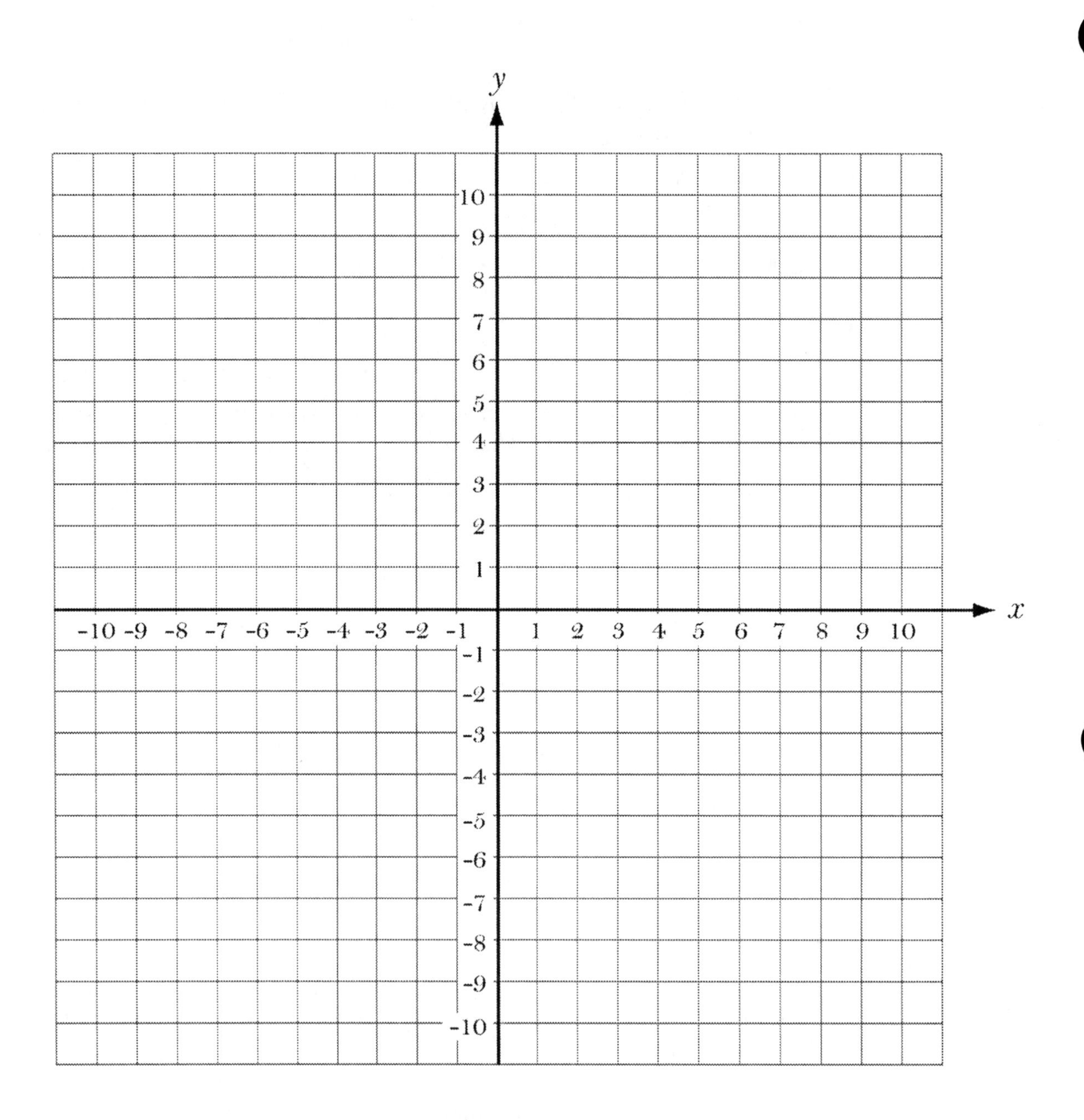

Conclusion	With practice, you should be able to graph the equation $y = mx + b$ quickly by using the slope and y-intercept. This activity helps you develop proficiency in applying these techniques.

Name Section Date

Activity 2.6 Change axes scales to change graph appearance.

Focus	Graphing linear equations, linear models
Time	15-20 minutes
Group size	2
Background	Often in business or politics people present information in the form of a graph to support their argument. In this activity you will see how the axes can be scaled on a graph to appear to support a desired argument.

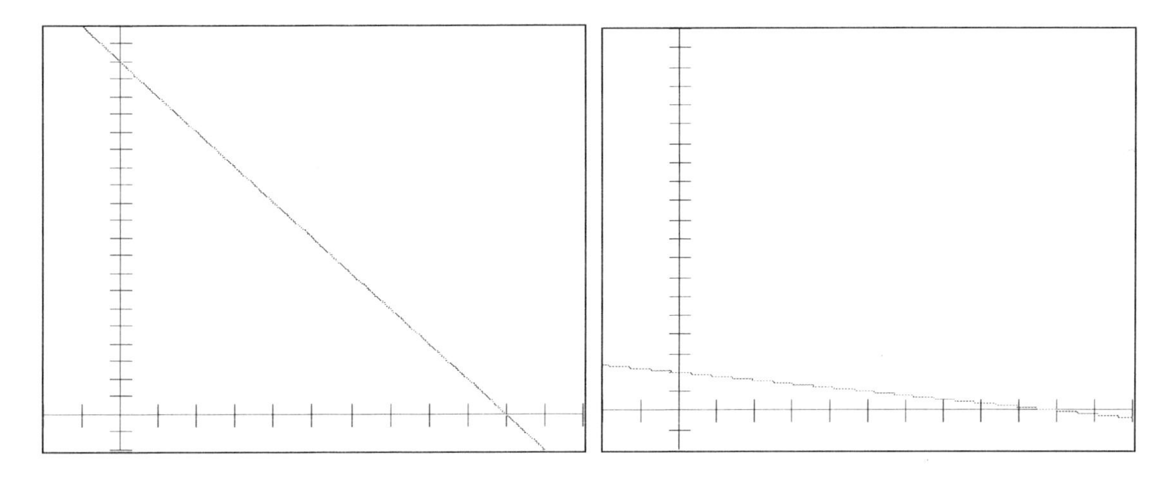

Look at the two graphs above. They look like they must represent different linear equations. However, both of these are graphs of the equation $y = -2x + 20$. They look very different because of the scales chosen for the *y*-axes. For the graph on the left, the *y*-axis goes from –2 to 20 with tick marks placed every 2 units. For the graph on the right, the *y*-axis goes from –20 to 200 with tick marks placed every 20 units. For both graphs the *x*-axes are the same (from –2 to 11 with tick marks placed every 1 unit). Normally, you wouldn't use different scales for your *x*- and *y*-axes unless the equation required it. However a person may choose to use differently scaled axes if they are attempting to have the data modeled by the equation appear to support their argument. Which graph do you think a person would choose if they wished to show that the data modeled by the equation $y = -2x + 20$ shows a very slow decline?

1. A company that sells high-end, customized SUV's sold 50 units in 1997. In 2005 that number had fallen to 25 units. Let x equal the number of years since 2000 and write the given data as two ordered pairs.

2. Work together to find the equation of a line that passes through these two ordered pairs.

3. Work together to graph this linear function on the axes below. Be sure to label the scales on your axes.

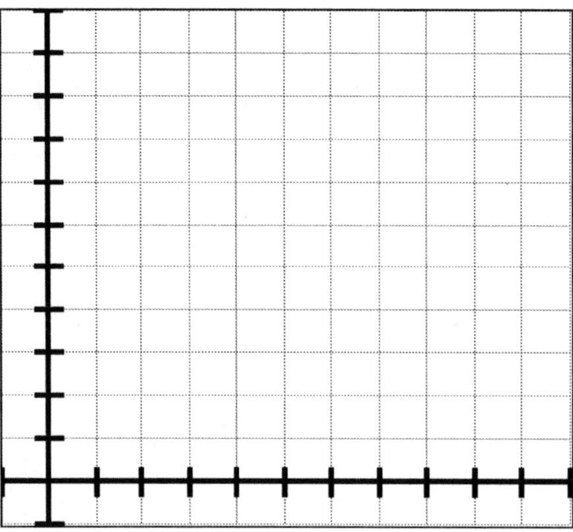

4. Choose one member of the group to be the current sales associate. This person would like to make the sales decline appear as slow and minimal as possible. The other member of the group will be a person interviewing to be the new sales associate. This person would like to show that the sales (under the current associate) are declining very rapidly. Each group member should now re-graph the equation, however each one should change the scales on the axes to make the graph appear to support their argument (hint: one may find it better to change the scale on the x-axis). Be sure to label the scales on your axes.

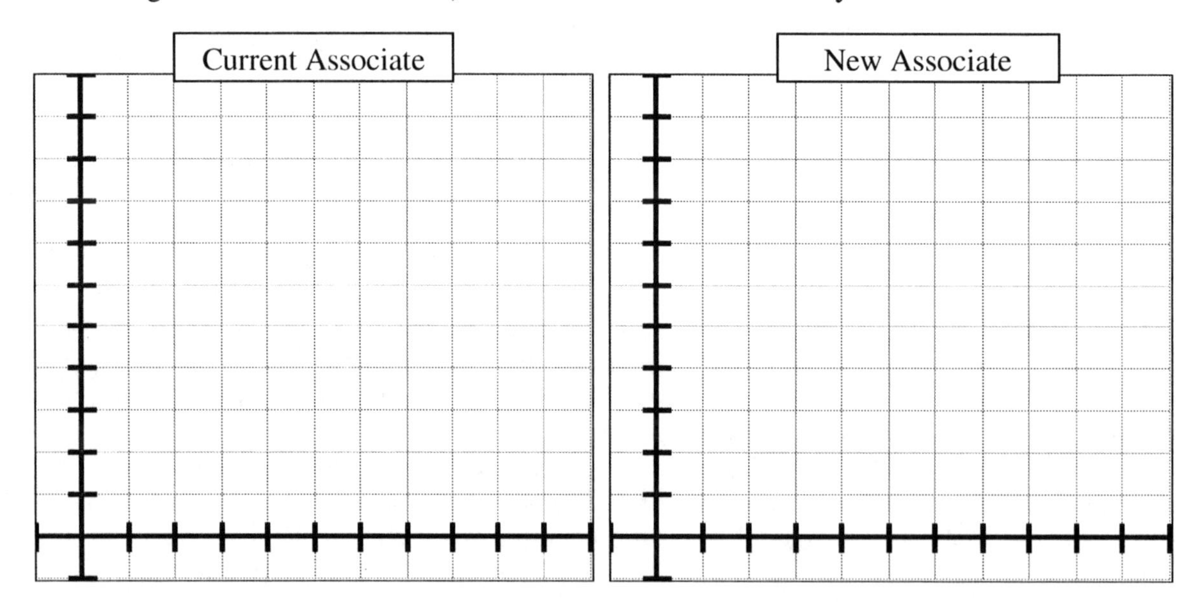

Conclusion	It is important to understand that manipulating the scales on the axes can make a graph appear somewhat different without changing the values on the graph itself. You should always label the scales on the axes when you are creating a graph and be aware of the scales when a graph is being presented to you. If the scales appear unusual or unnecessary, ask yourself the motives behind them.

Name	Section	Date

Sections 3.1, 3.2, and 3.3 Compare the three methods for solving systems of equations in two variables.

Focus	Solving systems of equations
Time	20–25 minutes
Group size	3
Materials	Ruler
Background	There are three methods for solving a system of two equations: graphing, substitution, and elimination. In this activity, all three methods will be used to solve each system. The solutions will be compared, and criteria will be established to help you decide which method is preferable for each system.

1. Tear out the three pages labeled Systems A, B, and C, and distribute one page to each group member. Begin by solving your system using the graphing method. Show your work on the page, and draw the graphs on the grid provided. Neatly organize and label your work, so the other group members can follow your steps.

2. When all group members are done, pass your paper to the group member on your left. Solve the system you receive using the substitution method. Show your work in the space provided. Again, be neat and organized so the other group members can follow your steps.

3. Finally, pass your paper to the group member on your left, and this time, solve the system using the elimination method.

4. When all group members are finished, study the three methods used to solve each system. Decide as a group what the preferred method or methods are for each system, and fill in the following table. Briefly state a reason for your group's choices.

System	Preferred method(s)	Reason
A		
B		
C		

5. Discuss your choices with another group. Do both groups agree on the preferred method(s) for solving each system? Decide whether there is a "right" and "wrong" method for solving any system.

Conclusion	Use the criteria you established in step 4 to help you decide which method to use when solving a system of two equations. While it may be easier to always rely on one method, it is a good idea to be versatile and use different methods as appropriate.

SYSTEM A $3x + 2y = 8,$ $5x - 3y = 7$

GRAPHING METHOD

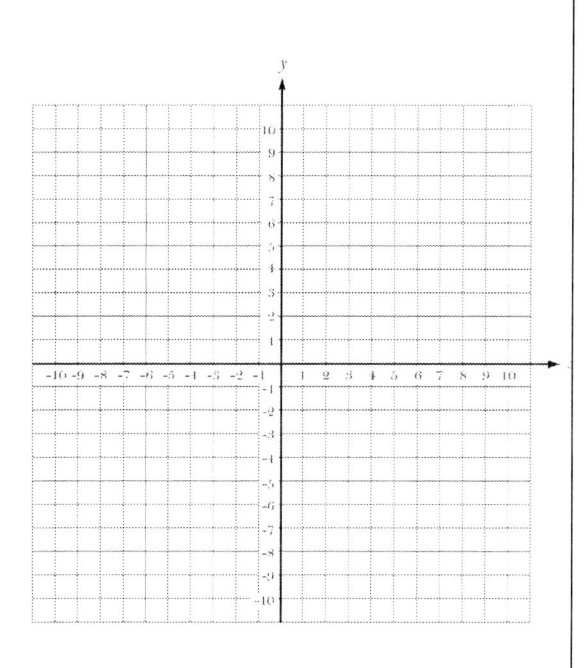

SUBSTITUTION METHOD	ELIMINATION METHOD

SYSTEM B

$$x + y = 4,$$
$$2x - y = 5$$

GRAPHING METHOD

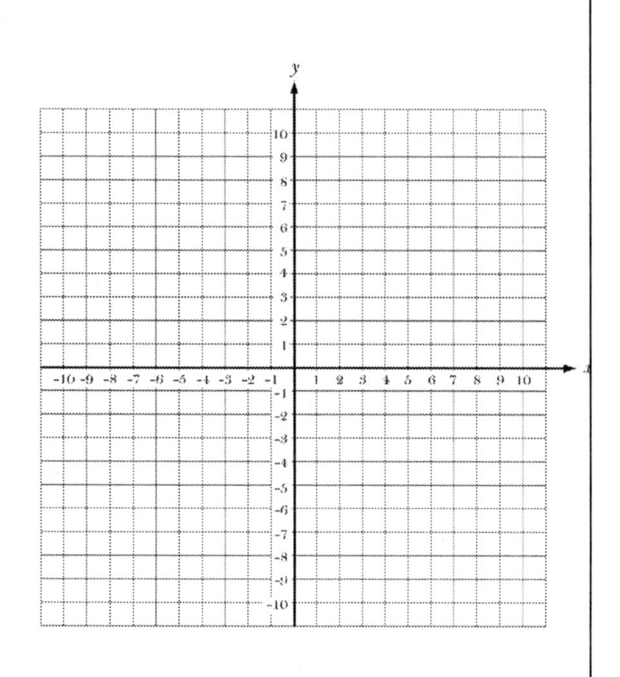

SUBSTITUTION METHOD

ELIMINATION METHOD

SYSTEM C

$$5x - y = 7,$$
$$3x + 2y = 12$$

GRAPHING METHOD

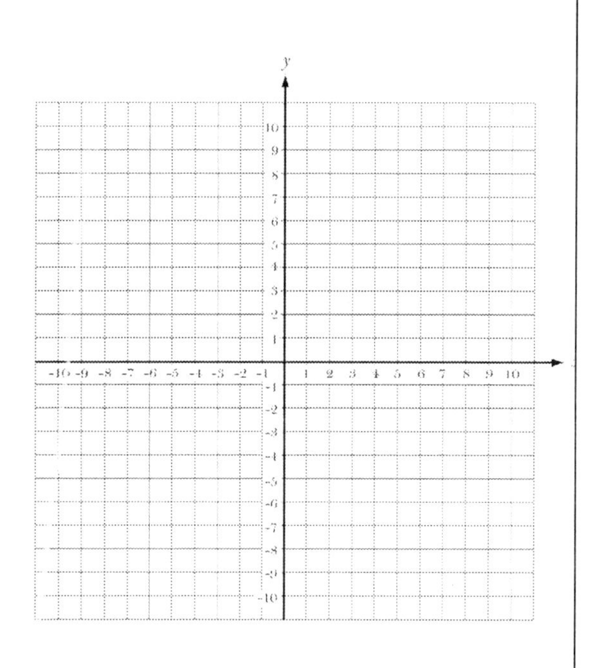

SUBSTITUTION METHOD

ELIMINATION METHOD

Name Section Date

Activity 3.5 Solve a system of equations in three variables by choosing a different variable to eliminate.

Focus	Systems of equations in three variables
Time	10–15 minutes
Group size	3
Background	When solving a system of equations in three variables, you need to choose a variable to eliminate. The ease of solving the system often depends on this choice of variable. In this activity, a system of equations will be solved three different ways, by choosing a different variable to eliminate first. By comparing the work for each method, you may gain some insight on how to choose the variable to eliminate first.

1. Review the six steps given in Section 3.5 of your textbook for solving systems of three equations. If necessary, refer to Examples 1–3 for clarification of the steps.

2. Working independently, each group member should solve the system given on the next page. One group member should begin by eliminating x, one should first eliminate y, and one should first eliminate z. Show your work in the space provided, and write neatly so that others can follow your steps.

3. Once each group member has solved the system, compare your answers. If the answers do not check, exchange papers and check each other's work. If a mistake is detected, allow the person who made the mistake to make the correction.

4. Decide as a group which of the three approaches (if any) ranks as easiest and which (if any) ranks as most difficult. Write your findings in the space below, using complete sentences. Then compare your rankings with the other groups in the class.

Conclusion	Now that you have some experience with choosing the variable to eliminate first, use your criteria whenever you need to solve a system of equations in three variables. You may find that the process can be simplified considerably depending on your choice.

$$10x + 6y + z = 7,$$
$$5x - 9y - 2z = 3,$$
$$15x - 12y + 2z = -5$$

Name Section Date

Activity 4.1 Determine the polynomial function for the number of handshakes possible in a group.

Focus	Polynomial functions
Time	20 minutes
Group size	5
Background	The number of possible handshakes within a group of n people can be given by a polynomial function $H(n)$. This activity leads you through the steps to find this function, and gives you practice in evaluating polynomial functions.

1. All group members should shake hands with each other. Without "double counting," how many handshakes occurred? Write your answer in the appropriate space in the table below.

2. Now, complete the following table. Form subgroups of 1, 2, 3, and 4, and count the number of handshakes in each case.

Group size	Number of Handshakes
1	
2	
3	
4	
5	

3. Try to find a function of the form $H(n) = an^2 + bn$, for which $H(n)$ is the number of different handshakes that are possible in a group of n people. Look for a pattern in the table to help you determine the values of a and b. Make sure $H(n)$ produces all of the values in the table above. Use the space below to show your work.

4. Compare the function found in the previous step with the function found by another group. Are they equivalent? If not, exchange ideas and try to come to a consensus. Use complete sentences in your answer.

5. Once both groups have agreed on the correct function, determine the number of handshakes for a group of size 10. If necessary, join with a third group so you have 10 in the group. First, use the function to predict the number of handshakes, then check your prediction by shaking hands with each other.

6. If there is time, and the class size is not too large, evaluate the function for the class size.

   H(class size) =

   Then, shake hands with every member of the class. Decide as a class how to organize this part of the activity so that each person shakes hands only once with the other members of the class. Count the number of handshakes, and verify that this number matches the value obtained from the handshake function.

Conclusion	The handshake function found in this activity is a good example of how functions are used in everyday life. When you encounter situations where there is a pattern to the data, see if you can find the function that fits the data.

Name Section Date

Activity 4.2 Derive the formulas for the squares of binomials.

Focus	Squares of binomials
Time	10–15 minutes
Group size	2
Background	Two binomials can always be multiplied using the distributive property. However, to multiply the square of a binomial, the special product formulas given in Section 4.2 of your textbook can be used. This activity shows you one method of deriving these formulas.

1. Multiply the following binomials using the distributive property. One group member should do the computations in Table 1, while the other group member does the computations in Table 2. Do your scratch work on a separate sheet of paper, and just write down the final product in the appropriate spaces in the table. The first problem has been done as an example.

Table 1

$(x+1)^2 = (x+1)(x+1) =$	x^2	+	2	x	+	1
$(x+2)^2 = (x+2)(x+2) =$	x^2	+		x	+	
$(x+3)^2 = (x+3)(x+3) =$	x^2	+		x	+	
$(x+4)^2 = (x+4)(x+4) =$	x^2	+		x	+	

Table 2

$(x-1)^2 = (x-1)(x-1) =$	x^2	−	2	x	+	1
$(x-2)^2 = (x-2)(x-2) =$	x^2	−		x	+	
$(x-3)^2 = (x-3)(x-3) =$	x^2	−		x	+	
$(x-4)^2 = (x-4)(x-4) =$	x^2	−		x	+	

2. Look at the results in your tables. Can you see a pattern in the coefficients of x and in the constant terms? In your own words, describe your observations.

3. Based on your observations, write down the product of the two binomials given below, without using the distributive property.

$$(x+7)(x+7) = \_\_\_\_\_\_\_\_\_\_\_\_\_\_\_\_\_\_\_\_$$

Now, notice that the multiplication above can be written as the square of a binomial. Complete the blanks below.

$$(x+7)(x+7) = (\qquad\qquad)^2 = \_\_\_\_\_\_\_\_\_\_\_\_\_\_\_\_\_\_\_\_$$

4. Finally, write down the final product of each of the following binomial squares, without using the distributive property. Follow the pattern from step 4.

$$(x+6)^2 =$$

$$(x-6)^2 =$$

$$(x-9)^2 =$$

5. Now, let's extend the results you have so far. Follow the procedure outlined in step 1, and multiply the binomials in Tables 3 and 4 using the distributive property. One group member should do the computations in Table 3, while the other group member does the computations in Table 4. Do your scratch work on a separate sheet of paper, and just write down the final answer in the appropriate spaces in the table.

Table 3

$(2x+1)^2 = (2x+1)(2x+1) =$	4	x^2	+	4	x	+	1
$(2x+3)^2 = (2x+3)(2x+3) =$		x^2	+		x	+	
$(3x+2)^2 = (3x+2)(3x+2) =$		x^2	+		x	+	
$(3x+4)^2 = (3x+4)(3x+4) =$		x^2	+		x	+	

Table 4

$(2x-1)^2 = (2x-1)(2x-1) =$	4	x^2	−	4	x	+	1
$(2x-3)^2 = (2x-3)(2x-3) =$		x^2	−		x	+	
$(3x-2)^2 = (3x-2)(3x-2) =$		x^2	−		x	+	
$(3x-4)^2 = (3x-4)(3x-4) =$		x^2	−		x	+	

6. Analyze the results in the tables. In your own words, write down the pattern that you observe.

Use the pattern to multiply the following binomials without using the distributive property.

$(2x+5)^2 =$

$(5x-6)^2 =$

$(4x-9)^2 =$

7. In the space below, write down the special product formulas from Section 4.2 in your textbook.

$(A+B)^2 =$
$(A-B)^2 =$

Compare these formulas with the pattern you wrote down in step 7. Can you see how the formulas are the algebraic equivalent of your observations?

Conclusion	This activity should give you a better understanding of the special product formulas for the square of a binomial. While you can always multiply binomials using the distributive property, the formulas provide a shortcut for the multiplication. They will also prove useful when you need to factor trinomials later in Chapter 4.

Name Section Date

Activity 4.8 Create and solve quadratic equations as a group.

Focus	Solving quadratic equations
Time	20–25 minutes
Group size	4
Background	Solving quadratic equations is an important skill in algebra. This activity will give you practice in using the principle of zero products to solve quadratic equations.

1. Each group will create and solve quadratic equations by following the steps outlined below. Study the example so you understand the mechanics of this process.

		Example
Step 1	The first group member thinks of two numbers and writes x = one number or x = another number, on a piece of notebook paper, and passes the paper to the second group member.	4; –3 $x = 4$ or $x = -3$
Step 2	The second group member creates two binomials that will give the solutions from step 1. Write the binomials as a product, and set the product equal to zero. Write the new equation below the first one, and pass the paper to the third group member.	$(x-4)(x+3)=0$
Step 3	The third group member multiplies the two binomials, writes the new equation below the second one, and folds over the paper so that only the last equation is shown. He or she then passes the paper to the fourth group member.	$x^2 - x - 12 = 0$
Step 4	The fourth group member writes the equation on his or her record sheet, and solves the equation. When done, the group unfolds the notebook paper and checks that the solution matches the equation written by the first group member. If it does not, the group should find out where the mistake occurred.	$x^2 - x - 12 = 0$ $(x-4)(x+3)=0$ $x-4=0$ or $x+3=0$ $x=4$ or $x=-3$

RECORD SHEET

ROUND 1 Equation:

Solution:

ROUND 2 Equation:

Solution:

ROUND 3 Equation:

Solution:

2. Now you are ready to create your own equations. In Round 1, use only positive integers in step 1.

   Each group member begins with step 1, writing down the equation on a piece of notebook paper. He or she then passes the paper to the group member on his or her right.

   Modify the equation you receive according to step 2, then pass the paper to your right again. Continue until you reach step 4; you should have an equation that looks like $ax^2 + bx + c = 0$. Write down this equation on the record sheet on the previous page. Then, solve and check the equation, following the directions for step 4.

   When all group members are done, examine all four equations in your group and correct any errors.

3. For Round 2, you may use negative integers to create the solutions in step 1. Start with a fresh sheet of notebook paper and pass the equation around the group as before.

4. Round 3 gets a little more complicated, as you may use fractions for the two solutions in step 1.

Conclusion	Notice that you are writing equivalent equations in steps 2 and 3. This is one of the fundamental concepts in equation solving, as explained in Chapter 1 of your textbook. This activity should give you a clearer understanding of how the principle of zero products works, and consequently increase your proficiency in solving quadratic equations.

Name Section Date

Activity 5.4 Simplify complex rational expressions.

Focus	Complex rational expressions, multiplication and division of rational expressions
Time	25 - 30 minutes
Group size	4
Background	Simplifying complex rational expressions can be challenging. This activity will help you simplify these expressions by working as a group to complete the process in steps. For this process we will use Method 2 in Section 5.4 of your textbook.

Follow the five-step process listed below for each of the complex rational expressions on the next page. After each expression has been simplified you should rotate which member of your group completes which step in the process for the next expression. When you have finished simplifying each of the expressions every group member should have had a turn to complete each of the steps in the process at least once.

1. The first member of your group should take a sheet of notebook paper and draw two lines vertically down the page dividing the page into three equal sections. This person then writes the original complex rational expression at the top of the middle section of the page and then hands the sheet to the second group member.

2. The second group member then takes the sheet and writes the numerator of the *original* complex rational expression at the top of the left hand section. This person then adds or subtracts as necessary to combine this expression into one single rational expression and then hands the sheet to the third group member.

3. The third group member then takes the sheet and writes the denominator of the *original* complex rational expression at the top of the right hand section. This person then adds or subtracts as necessary to combine this expression into one single rational expression and then hands the sheet to the fourth group member.

4. The fourth group member then takes the sheet and rewrites the *original* expression in the middle as a division problem. This will be the division of two fractions:
   (result from the left hand side of page) $\div$ (result from the right hand side of page).
   The sheet should be handed back to the first group member.

5. The first group member takes the sheet and performs the division in the middle of the sheet using the methods discussed in Section 5.1 of the textbook. Be sure that the final answer is simplified by removing any common factors.

Example:

Step 2: Numerator Work	**Step 1:** Original Problem	**Step 3:** Denominator Work
$\frac{8}{x^2-9}-\frac{5}{x+3}$ $=\frac{8}{(x+3)(x-3)}-\frac{5}{x+3}$ $=\frac{8}{(x+3)(x-3)}-\frac{5}{(x+3)}\bullet\frac{x-3}{x-3}$ $=\frac{8}{(x+3)(x-3)}-\frac{5(x-3)}{(x+3)(x-3)}$ $=\frac{8-5(x-3)}{(x+3)(x-3)}$ $=\frac{8-5x+15}{(x+3)(x-3)}$ $=\frac{-5x+23}{(x+3)(x-3)}$	$\frac{\frac{8}{x^2-9}-\frac{5}{x+3}}{\frac{3}{x^2-9}-\frac{4}{x-3}}$ **Step 4:** Rewrite Original $\frac{-5x+23}{(x+3)(x-3)}\div\frac{-4x-9}{(x+3)(x-3)}$ **Step 5:** Divide & Simplify $\frac{-5x+23}{(x+3)(x-3)}\div\frac{-4x-9}{(x+3)(x-3)}$ $=\frac{-5x+23}{\cancel{(x+3)(x-3)}}\bullet\frac{\cancel{(x+3)(x-3)}}{-4x-9}$ $=\frac{-5x+23}{-4x-9}$ Final Answer: $\frac{-5x+23}{-4x-9}$	$\frac{3}{x^2-9}-\frac{4}{x-3}$ $=\frac{3}{(x+3)(x-3)}-\frac{4}{x-3}$ $=\frac{3}{(x+3)(x-3)}-\frac{4}{x-3}\bullet\frac{x+3}{x+3}$ $=\frac{3}{(x+3)(x-3)}-\frac{4(x+3)}{(x+3)(x-3)}$ $=\frac{3-(4x+12)}{(x+3)(x-3)}$ $=\frac{3-4x-12}{(x+3)(x-3)}$ $=\frac{-4x-9}{(x+3)(x-3)}$

Group Problems:

$$\frac{\frac{1}{x}-\frac{1}{y}}{\frac{1}{x^3}-\frac{1}{y^3}} \qquad \frac{\frac{2}{x+3}+\frac{5}{x-1}}{\frac{10}{x-1}-\frac{3}{x+3}} \qquad \frac{\frac{1}{x^2-1}+\frac{3}{x+1}}{\frac{7}{x^2-1}-\frac{5}{x-1}} \qquad \frac{\frac{1}{x^2+3x+2}+\frac{2}{x^2-2x-3}}{\frac{5}{x^2-3x-4}-\frac{1}{x^2-2x-8}}$$

Conclusion	Simplifying complex rational expressions can be a lengthy process. It can help to break the process down into smaller, well-organized steps.

Name Section Date

Activity 5.7 Develop a formula for calculating the time required to complete a task when working together.

Focus	Formulas, work problems
Time	15–20 minutes
Group size	2
Background	Solving formulas is one of the more useful algebraic skills. This activity will lead you through an intuitive development of a formula for calculating the time required for two people to complete a task working together.

1. Use the method shown in Section 5.6 of your textbook to calculate the time, in hours, required to complete a task working together, for the individual times given in the table below. Leave your answers in fractional notation.

   Split the work with your partner, so each does two problems.

Time required for first worker (hours)	Time required for second worker (hours)	Time required working together (hours)
2	3	
2	5	
3	4	
3	5	

2. Study the numerator of each answer. Can you see a connection between the individual times and the numerator? State this relationship in your own words.

   Now study the denominator of each answer. What is the relationship between the individual times and the denominator?

In your own words, complete the formula below, using the relationships you wrote for the numerator and denominator.

Time required working together = ________________________________

3. Use the formula above to solve the following problem.

   How long would it take two people to complete a task if each person takes 4 hours and 5 hours to do the job individually?

Test your formula by using the method shown in the textbook to solve the problem. If necessary, modify your formula so the answers match.

4. Now, let's write the formula using algebraic symbols. Suppose one person takes a hours to complete a task alone, and a second person takes b hours to complete the same task alone. Let t represent the time it would take both people to complete the task working together. Follow the formula developed in step 2, and write an algebraic formula for the time required working together.

t = ____________________

5. From Section 5.6 of your textbook, the equation used to solve work problems is

$$\frac{t}{a} + \frac{t}{b} = 1$$

Using the methods shown in Section 5.7 of your textbook, solve the equation for the variable t.

Compare your solution with the formula you developed in step 4. Are they identical? Discuss any discrepancies with your partner. If necessary, check your work with that of another group, and try to resolve any differences.

Conclusion	You can use the work formula developed in this activity to solve work problems from Section 5.6 of your textbook. The skills developed here can also be used when solving other formulas in your textbook.

Name Section Date

Activity 5.8 Model the height of a bouncing ball with an equation of direct variation.

Focus	Direct variation
Time	25 minutes
Group size	3
Materials	Rubber ball or tennis ball, tape measure, ruler
Background	Does a rubber ball always rebound a fixed percentage of the height from which it is dropped? In other words, does the rebound height vary directly as the height from which the ball is dropped? To answer this question, each group will perform an experiment, and gather data, and try to find the equation of direct variation that models this situation.

NOTE: Read all the steps before beginning this activity.

1. One group member should hold a rubber ball at some height above the floor. A second group member should measure this height. The ball should then be dropped and caught at the peak of its bounce. The second group member should measure this rebound height and the third group member should record the measurements in the table below. Repeat this procedure two more times from the same height. Then find the average of the three rebound heights, and write the value in the table.

	Rebound height			
Original height	First try	Second try	Third try	Average

2. Repeat step 1 four more times at different heights, and record the data in the table above. The five data points form five ordered pairs in the form (original height, average rebound height).

3. Graph the five data points on the grid below, and draw a straight line starting at (0, 0), that comes as close as possible to all the points.

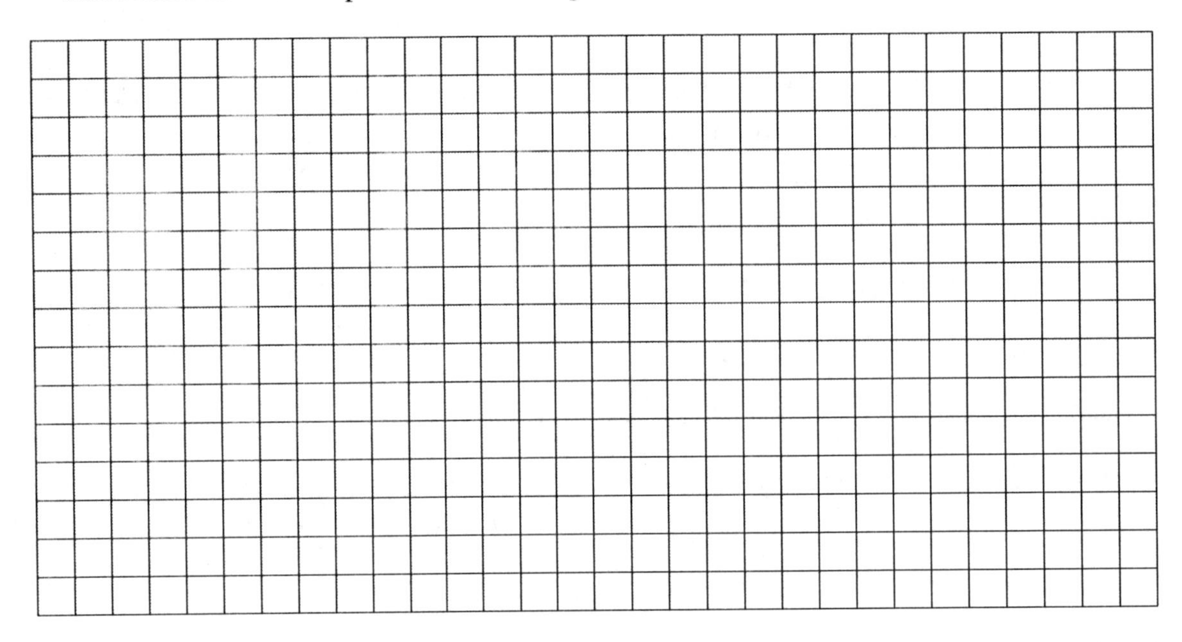

4. Follow the procedure shown in Example 1 in Section 5.8 of your textbook to find the equation of variation for the line. Choose one of the ordered pairs, and substitute into the equation $y = kx$. Show your work in the space below.

Equation of variation: $y =$ __________ x

5. Use the equation of variation from the previous step to predict the rebound height for a ball dropped from a height that is 2 ft more than the greatest height used in steps 1 and 2.

Original height = __________

Predicted rebound height = __________

Actual rebound height = __________

How close is the predicted rebound height to the actual rebound height? What factors could explain any difference? Use complete sentences in your answer.

6. <u>Technology Extra:</u> Use the linear regression feature on your graphing calculator to fit a linear equation to the data. Refer to the Calculator Spotlight in Section 2.6 of your textbook, if needed. Then use that equation to complete the other exercises. Discuss which equation is better.

Conclusion	As you observed in this activity, the rebound height of a rubber ball varies directly as the height from which the ball is dropped. When the data is plotted on a graph, you can see that the variation constant is also the slope of the line.

Name Section Date

Activity 6.1 Use the skid length function to determine a safe following distance when driving behind another vehicle.

Focus	Square root functions
Time	15–25 minutes
Group size	3
Materials	Calculator
Background	The faster a car is traveling, the more distance it needs to stop. Thus it is important for drivers to allow sufficient space between their vehicle and the vehicle in front of them. Police recommend that for each 10 mph of speed, a driver allow 1 car length of following distance. Thus a driver traveling at 30 mph should have at least 3 car lengths between his or her vehicle and the one in front. In this activity, we will use the skid length function in Exercise 29 in Exercise Set 6.1 to determine the safe following distance for various cars.

The function $S(x) = 2\sqrt{5x}$ can be used to estimate the speed at which a car was traveling. $S(x)$ is the speed, in miles per hour, and x is the length of the skid mark, in feet.

1. Each group member should estimate the length of a car in which he or she frequently travels. Using a calculator as needed, each group member should complete the table on the top of the next page. Follow the directions below to calculate the values for the second and third columns.

   The recommended following distance is 1 car length for each 10 miles per hour, as specified by police.

   Suppose you are travelling at 50 mph, and your car length is 20 feet.

   The recommended following distance will be 5(20) = 100 feet.

   The skid length speed is the speed a vehicle would have to travel to produce a skid length equal to the recommended following distance.

   For the example given, the skid length speed is

$$S(100) = 2\sqrt{5 \cdot 100} = 2\sqrt{500} \approx 45 \text{ mph}$$

Car length ______________________

Speed (mph)	Recommended following distance (ft)	Skid length speed, $S(x)$ (mph)
30		
40		
50		
60		
70		

2. When you are done, examine your table. Determine, as a group, whether there are any speeds at which the recommended following distance of "1 car length per 10 mph" might not be sufficient. On what reasoning do you base your answer? Use complete sentences in your answer.

3. Compare tables with the others in your group. Determine, as a group, how car length affects the results. Would you recommend the "1 car length per 10 mph" rule to a new driver? Explain your answer using complete sentences.

4. Some driver education books recommend a following distance of 2 seconds. Thus, if the car in front of you passes a marker, you should pass the same marker a minimum of 2 seconds later. With this rule of thumb, the length of your car is not important. As a group, complete the table below.

   To find the recommended following distance, you would need to convert the units from mph to ft/s. Use the conversion factor of 88/60. Then multiply the speed by 2 seconds to get the following distance.

   For a speed of 30 mph, the following distance is

   $$30 \cdot (88/60) \cdot 2 = 88 \text{ ft.}$$

   The skid length speed is calculated the same way as before.

   $$S(88) = 2\sqrt{5 \cdot 88} = 2\sqrt{440} \approx 42 \text{ mph}$$

Speed (mph)	Recommended following distance (ft)	Skid length speed, $S(x)$ (mph)
30	88	42
40		
50		
60		
70		

5. As before, determine, as a group, whether there are any speeds at which the recommended following distance of "2 seconds" might not be sufficient. On what reasoning do you base your answer? Use complete sentences in your answer.

6. Would you recommend the "2 second" rule to a new driver? Explain your answer using complete sentences.

7. If it has just rained, how would you change your recommendations? Use complete sentences in your answer.

Conclusion	By using mathematical reasoning, you were able to evaluate the accuracy of two rules of thumb for following distances. There are other situations that arise in everyday life where mathematics may be able to provide you with insight into accepted recommendations. Use the techniques learned in this activity to analyze these rules and check for yourself whether they are accurate.

Name Section Date

Activity 6.2 Investigate the effect of the order of rational exponents on exponential functions.

Focus	Rational exponents, functions
Time	10–20 minutes
Group size	3
Materials	Calculator
Background	In arithmetic, $\frac{3}{5}$, $\frac{1}{10}\cdot 6$, and $6\cdot\frac{1}{10}$ all represent the same number. However, the functions $f(x)=x^{3/5}$, $g(x)=\left(x^{1/10}\right)^6$, and $h(x)=\left(x^6\right)^{1/10}$ each represent different functions. In this activity, you will analyze the graphs of these functions, and discover their differences, if any. Furthermore, the definition of $a^{m/n}$ will be used to help you explain why these differences exist.

1. Each group member should choose one of the functions given on the next page. Complete the table of values, and graph the function on the grid provided. Be sure that you pay close attention to the definition of $a^{m/n}$ given in Section 6.2 of your textbook. In particular, remember that the even root of a negative number is not a real number.

2. Compare the three graphs and check each other's work. How and why do the graphs differ? Use complete sentences in your answer.

3. Decide as a group which graph, if any, would best represent the graph of $k(x)=x^{6/10}$. Then, explain your reasoning to the entire class.

4. Technology Extra: Repeat Steps 1–3 using a graphing calculator. Do the results change?

Conclusion	As you saw, the definition of $a^{m/n}$ creates restrictions on the domains of the functions depending on whether the power or root occurs first. Thus, even though the quantities $\frac{1}{10}\cdot 6$, and $6\cdot\frac{1}{10}$ are equivalent, the position of the factors in the exponent of a function is very important.

$f(x) = x^{3/5}$

x	$f(x)$
0	
1	
3	
5	
–1	
–3	
–5	

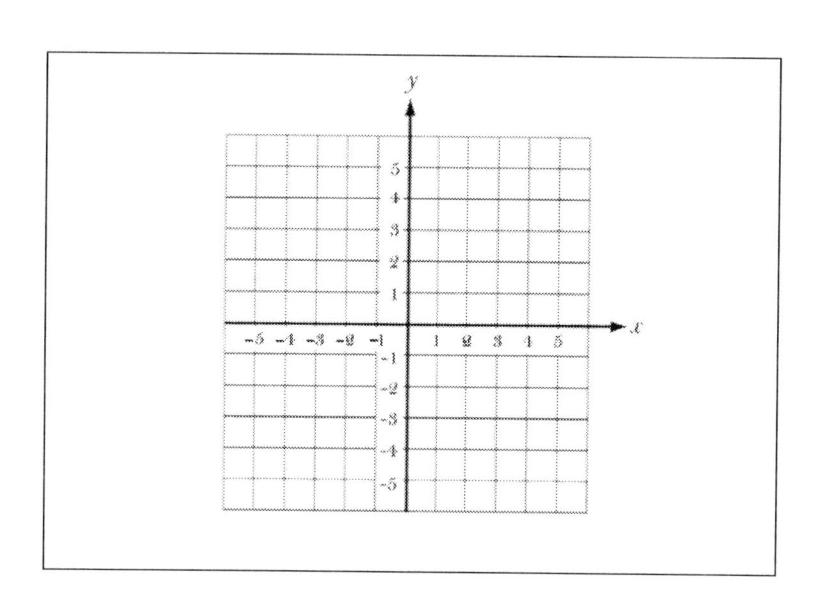

$g(x) = \left(x^{1/10}\right)^6$

x	$g(x)$
0	
1	
3	
5	
–1	
–3	
–5	

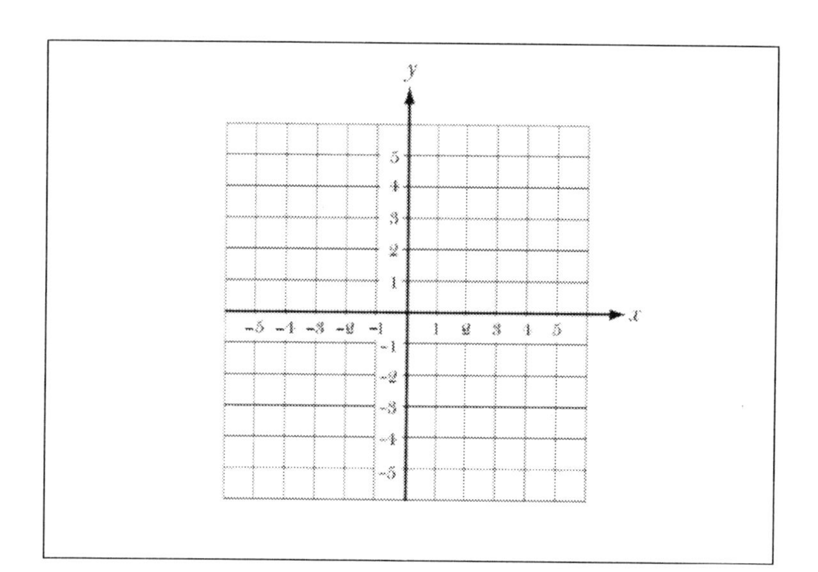

$h(x) = \left(x^6\right)^{1/10}$

x	$h(x)$
0	
1	
3	
5	
–1	
–3	
–5	

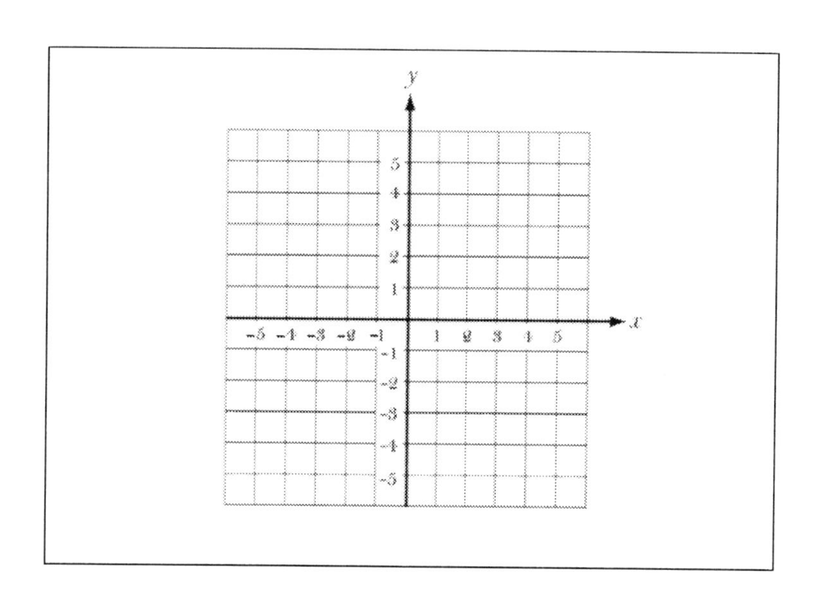

Name Section Date

Activity 6.6 Develop a formula for the swing time of a pendulum.

Focus	Square roots, formulas
Time	25–30 minutes
Group size	3
Materials	Tape measure, watch or clock to measure time in seconds, pendulum (see below), calculator
Background	A pendulum is simply a string, a rope, or a chain with a weight of some sort attached at one end. When the unweighted end is held, a pendulum can swing freely from side to side. A shoe hanging from a shoelace, a yo-yo, a pendant hanging from a chain, a fishing weight hanging from a fish line, or a hair brush tied to a length of dental floss are all examples of a pendulum. In this activity, each group will develop a mathematical model or formula that relates a pendulum's length L to the time T that it takes for one complete swing back and forth (one "cycle").

1. Make a pendulum by attaching a weight to one end of a string. The size of the weight is unimportant; it should be heavy enough so the string is pulled tight and the pendulum swings smoothly, but not so heavy that the string would break.

2. One group member holds the string so that its length is 1 foot. A second group member should lift the weight to one side and then release it; do not throw the weight. It does not matter how high you lift the weight. The third group member should determine the time, in seconds, for one complete swing (cycle), by timing five cycles and dividing the time by 5. Record this time in the table below.

   Repeat this procedure for each length listed in the table.

Length, L (in feet)	Time, T (in seconds)
1	
1.5	
2	
2.5	
3	
3.5	

3. Examine the table your group has created. Can you find a number, a, such that $T \approx aL$ for all pairs of values in the table?

$$T \approx \_\_\_\_\_\_\_\_\_\_ \cdot L$$

4. To see if a better model can be found, add a third column to the table and fill in $\sqrt{L}$ for each value of L listed. Copy the times from the table in step 1.

L	$\sqrt{L}$	T
1		
1.5		
2		
2.5		
3		
3.5		

Can you find a number, b, such that $T \approx bL$? Does this appear to be a more accurate model than $T \approx aL$?

$$T \approx \_\_\_\_\_\_\_\_\_\_ \cdot L$$

5. Use the model from step 3 to predict the time for one complete cycle when the string is 4 feet long. Then check your prediction by measuring T as you did in step 1. Was your prediction "acceptable"? Compare your results with those of another group.

6. In the Exercise Set for Section 6.6 of your textbook, there is a formula relating T and L, as stated below.

$$T = 2\pi\sqrt{\frac{L}{32}} = \frac{2\pi}{\sqrt{32}} \cdot \sqrt{L}$$

Use a calculator to figure out the decimal equivalent of $2\pi / \sqrt{32}$. Rewrite the above formula using this number.

$$T \approx \_\_\_\_\_\_\_\_\_\_ \cdot L$$

How does the value of b from step 3 compare with the decimal equivalent of $2\pi / \sqrt{32}$?

Conclusion	The formula you developed that relates the pendulum's length to the swing time is known as an empirical formula, since it was derived from experimental data. The theoretical formula in step 5 is obtained by applying physics principles to the situation. Both approaches are important in the study of physics.

Name	Section	Date

Activity 7.1 Discover the rule for completing the square.

Focus	Completing the square
Time	15–20 minutes
Group size	2
Background	The process of completing the square is used to solve any quadratic equation in Section 7.1 of your textbook. This activity leads you through a discovery process to understand the rule for completing the square.

1. Multiply the following binomials using either the distributive property or the special products for the square of a binomial (Section 4.2). One group member should do the computations in Table 1, while the other group member does the computations in Table 2. Do your scratch work on a separate sheet of paper, and just write down the final products in the appropriate spaces in the table.

Table 1

$(x+1)^2 =$	x^2	+		x	+	
$(x+2)^2 =$	x^2	+		x	+	
$(x+3)^2 =$	x^2	+		x	+	
$(x+4)^2 =$	x^2	+		x	+	

Table 2

$(x-1)^2 =$	x^2	−		x	+	
$(x-2)^2 =$	x^2	−		x	+	
$(x-3)^2 =$	x^2	−		x	+	
$(x-4)^2 =$	x^2	−		x	+	

2. Look at the results in your tables. Compare the constant term in the binomial with the coefficient of x in the trinomial. Write your observations below, using complete sentences.

Then compare the constant term in the binomial with the constant term in the trinomial. As before, write your observations using complete sentences.

Finally, compare the coefficient of x in the trinomial with the constant term in the trinomial. Use your own words to describe the pattern observed. Write your answer in complete sentences.

3. Now, let's try going the other way. Use your observations from the previous step to help you complete the tables below. One group member should do the computations in Table 3, while the other group member does the computations in Table 4.

Table 3

$x^2 + 8x + 16 =$	$(\qquad)^2$
$x^2 + 10x + 25 =$	$(\qquad)^2$
$x^2 + 12x +$ ______ $=$	$(\qquad)^2$
$x^2 + 14x +$ ______ $=$	$(\qquad)^2$
$x^2 + 16x +$ ______ $=$	$(\qquad)^2$

Table 4

$x^2 - 8x + 16 =$	$(\qquad)^2$
$x^2 - 10x + 25 =$	$(\qquad)^2$
$x^2 - 12x +$ ______ $=$	$(\qquad)^2$
$x^2 - 14x +$ ______ $=$	$(\qquad)^2$
$x^2 - 16x +$ ______ $=$	$(\qquad)^2$

4. When you are both done, look at the results in the tables. In your own words, explain how you found the constant term of the trinomial in the last 3 exercises in each table. Use complete sentences.

5. From Section 7.1 of your textbook, the rule for completing the square is:

   When solving an equation, to complete the square of an expression like $x^2 + bx$, we take half the coefficient of the x-term, which is $b/2$, and square. Then we add that number, $(b/2)^2$ on both sides.

   Does this rule match the one you wrote in step 4?

Conclusion	This activity should give you a better understanding of the process of completing the square. Instead of memorizing a rule, you now have a procedure for finding the constant term of the perfect square trinomial. Many rules in algebra are the result of an observed pattern. Whenever you encounter such rules, try to find out what the underlying pattern is and you may increase your understanding of the rule.

Name Section Date

Activity 7.5 Practice graphing and identifying the graphs of quadratic functions.

Focus	Graphs of quadratic functions
Time	15–20 minutes
Group size	4
Background	In Section 7.5 of your textbook, you saw that the graph of the quadratic function $f(x)=a(x-h)^2+k$ looks like that of $y=ax^2$ but shifted horizontally (right or left), and vertically (up or down), depending on the values of h and k. In this activity, you will practice graphing functions by shifting the graph horizontally and/or vertically.

1. On each of four scraps of paper, write one of the following functions.

$$f(x)=\frac{1}{2}(x+3)^2+1$$

$$g(x)=\frac{1}{2}(x+3)^2-1$$

$$h(x)=\frac{1}{2}(x-3)^2+1$$

$$k(x)=\frac{1}{2}(x-3)^2-1$$

2. Place the scraps of paper face down on the desk, and mix them up. Then, one by one, each group member should select one of the pieces of paper. Do not show your function to the other members of your group.

3. Each group member should carefully graph the function selected using the grid on the next page. Use the techniques of shifting as described in Section 7.5 of your textbook to do your graph. Make sure the graph is neat, but do not label the graph with the function.

4. When all group members have finished graphing their function, place the graphs in a pile. The group should then agree on the correct function for each graph *with no help from the person who drew the graph.* Again, use the techniques of shifting to help you match the graphs with the functions. Write the function on the graph.

5. Compare your group's labeled graphs with those of other groups to reach consensus within the class on the correct label for each graph.

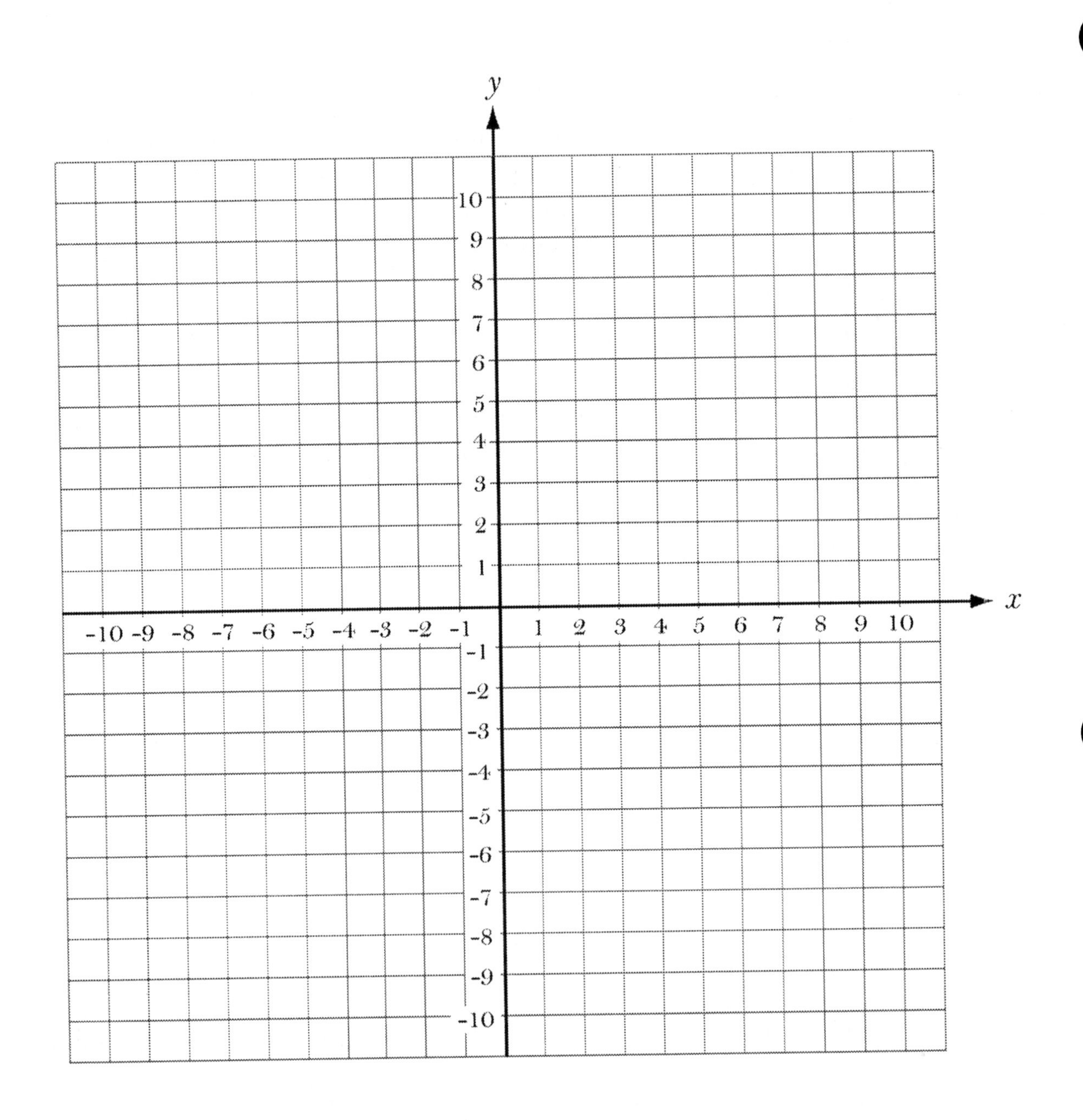

Conclusion	With practice, you should be able to graph the function $f(x) = a(x-h)^2 + k$ by starting with the graph of $y = ax^2$ shifting it up or down, and left or right. This activity helps you develop proficiency in applying these techniques.

Name Section Date

Activity 7.7 Fit a quadratic function to a set of data.

Focus	Modeling quadratic functions
Time	20–30 minutes
Group size	4
Materials	Calculator
Background	There are many situations for which data can be modeled by a quadratic function. This activity gives you practice finding a quadratic function to fit the given data.

1. A vintage stereo system has a counter for finding locations on an audio cassette. When a fully wound cassette with 45 min of music on a side begins to play, the counter is at 0. After 15 min of music has played, the counter reads 250 and after 35 min, it reads 487. When the 45-min side is finished playing, the counter reads 590. The data points are summarized in the table below. Note that the counter numbers are divided by 100 to make the computations easier.

Counter number	Minutes played
0	0
2.50	15
4.87	35
5.90	45

2. Use the data points to find a quadratic function that fits the data. Since you only need three ordered pairs to find the function, each group member should select a different set of three points from the four given. Write your function in the form

$$T(n) = an^2 + bn + c,$$

where $T(n)$ represents the time, in minutes, that the tape has run at counter reading n hundred. Show your work in the space provided on the next page. Write neatly so that another group member can follow your steps.

3. Exchange papers with another group member, and check the work on the paper you receive. If you find any errors, discuss them with the person who did the work. He or she should then correct the errors. When you are satisfied with the accuracy of the work, return the paper to the owner.

4. Now, let's determine which function best fits the data. Evaluate your function for the counter reading that was not used to find the function. Show your work below.

   Does your function come close to predicting the number of minutes played at the counter reading used? Compare your prediction with the other group members' predictions. Which function comes closest to predicting the minutes played? Use complete sentences in your answer.

5. The same counter used above reads 432 after the tape has played for 30 min. Evaluate your function at $n = 4.32$. Compare your result with the other group members. Which function comes closest to predicting this data point?

6. <u>Technology Extra:</u> Use a graphing calculator to fit a quadratic function to the data. Refer to the Calculator Spotlight in Section 7.7 of your textbook, if needed. Note that all four data points can be used. Then complete the other steps of this activity. Compare this function with the others found using 3 data points. Which seems best?

Conclusion	As you saw in this activity, data analysis can lead to different answers even though the procedures are correct. This occurs because different sets of data points lead to slightly different quadratic functions. In fact, the more data points you use to find the function, the more accurate the predictions.

Name Section Date

Activity 8.2 Create composite functions and deduce the original functions.

Focus	Composite functions
Time	10–15 minutes
Group size	2
Background	The ability to find composite functions from two given functions is important in the study of algebra. Conversely, there are situations in calculus where it is important to recognize how a function can be expressed as a composition. This activity gives you practice with both these processes.

1. Each group member should create two functions. Do not let your partner see your functions! Use the functions in Exercises 33–40 in Exercise Set 8.2 of your textbook as a guide when writing your functions.

   $f(x)$ = ________________________

   $g(x)$ = ________________________

2. Follow the steps given in Example 10 in Section 8.2 of your textbook to find the composite function values $f \circ g(2)$ and $f \circ g(-3)$. Then find the composite function $f \circ g(x)$. Show your work below.

   $f \circ g(x)$ = __________________________________

   Tear out the next page and copy the composite function in the space provided. Do not write the original functions.

3. Exchange papers with your partner. On the paper you receive, try and find the original functions $f(x)$ and $g(x)$ that were used to obtain the composite function. Write your guesses in the appropriate space, and check by forming the composition.

4. Exchange papers again, and compare the guessed functions with the functions that you started out with. If they do not match, exchange papers, and let your partner guess different functions. Continue until he or she guesses the same functions that you started out with.

5. If you have time, repeat steps 1 to 4, with different functions. Try to create more complicated functions, so the process of finding the original functions is more challenging.

Composite function, $f \circ g(x)$ = ______________________________

Guesses: $f(x)$ = ____________________

$g(x)$ = ____________________

Check:

Conclusion	As stated in Example 12 in Section 8.2 of your textbook, there may be several answers when finding the original functions that were used to get the composite function. You may also have discovered this in the process of completing the activity.

Name Section Date

Activity 8.3 Practice graphing logarithmic functions and their inverses.

Focus	Graphs of logarithmic functions
Time	15–20 minutes
Group size	3
Materials	Calculator
Background	The logarithmic function $f^{-1}(x) = \log_a x$ is defined in Section 8.3 of your textbook as the inverse of the exponential function $f(x) = a^x$. Thus, the graphs of these functions are reflections of each other across the line $y = x$. This activity gives you practice graphing both functions and verifying that they are indeed inverses.

1. On each of three scraps of paper, write one of the following functions.

$$f^{-1}(x) = \log_2 x \qquad g^{-1}(x) = \log_3 x \qquad h^{-1}(x) = \log_4 x$$

2. Place the scraps of paper face down on the desk, and mix them up. Then, one by one, each group member should select one of the pieces of paper. Do not show your function to the other members of your group.

3. Each group member should carefully graph the function selected using the grid on the next page. Use the method shown in Example 1 in Section 8.3 of your textbook to draw your graph. Make sure the graph is neat, but do not label the graph with the function.

4. When all group members have finished graphing their function, pass your paper to the group member on your left. On the paper you receive, draw the inverse function by reflecting the graph across the line $y = x$. Refer to Examples 8 and 9 in Section 8.2, as needed.

5. When all group members are done, pass your paper to the left again. Examine the graphs you receive, and guess the functions for each graph. Write your guesses on the paper, and pass the paper to the left. You should receive your original paper back.

6. Compare the guessed logarithmic function with your original function. If they do not match, discuss this with the person who made the guess. Then check that the exponential function is the correct inverse of the logarithmic function. Again, discuss any differences with the group member who made the guesses.

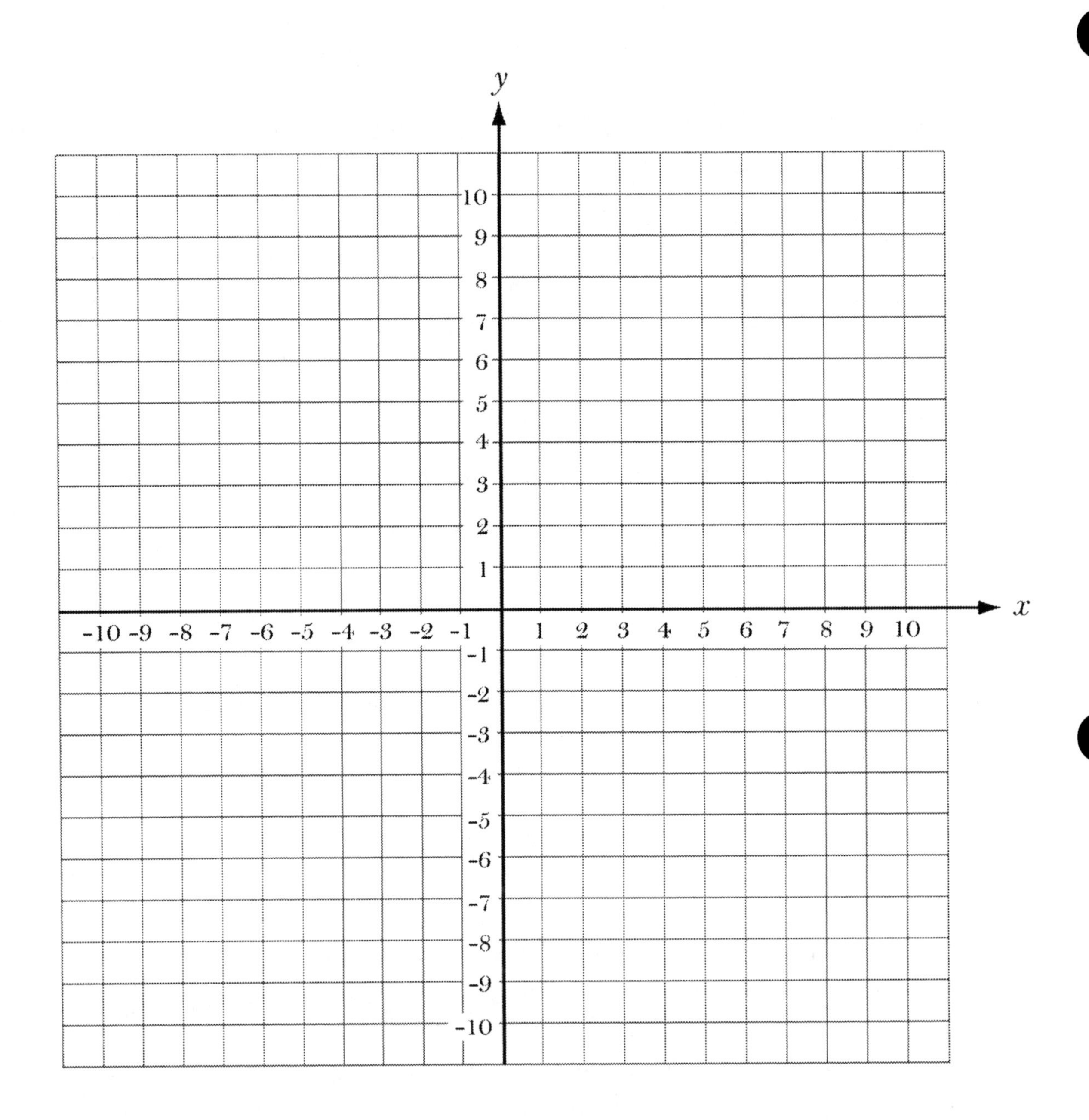

Conclusion	This activity should help you understand the relationship between logarithmic and exponential functions. Remember that the logarithm of a number is an exponent, thus converting from exponential form to logarithmic form is simply a matter of finding the inverse.

Name Section Date

Activity 8.6 Analyze the doubling time of money invested in an account paying compound interest.

Focus	Solving exponential equations, using formulas
Time	20 - 25 minutes
Group size	3
Background	Exponential equations have wide applications in real-life. One specific application is the analysis of doubling time for money invested in an account paying compound interest. Using your ability to solve exponential equations you will be able to see the impact different values in the compound interest formula have on the final result.

The formula for calculating the amount in an account paying compounded interest after a period of time is given in Section 8.1 of the text. In this formula P represents the amount initially invested (the principal), r represents the interest rate (expressed as a decimal), n represents the number of compounding sessions per year, t represents the number of years, and A represents the amount in the account at the end of the time period. In this activity you will use both this formula and the methods for solving exponential equations discussed in Section 8.6 of the text.

$$A = P\left(1 + \frac{r}{n}\right)^{nt}$$

1. Each member of your group should answer one of the questions listed below (you will need a calculator).

   How long (in years) does it take $4,000.00 to double to $8,000.00 if it is invested in an account paying 8.0% interest compounded quarterly (four times per year)?

   How long (in years) does it take $700.00 to double to $1,400.00 if it is invested in an account paying 8.0% interest compounded quarterly (four times per year)?

   How long (in years) does it take $50.00 to double to $100.00 if it is invested in an account paying 8.0% interest compounded quarterly (four times per year)?

2. Compare results with your group. Does it appear that the amount initially invested in the account (principal) has an impact on the length of time it takes to double that amount (if the interest rate and compounding period remain constant)?

3. Each member of your group should answer one of the questions listed below (you will need a calculator).

    How long (in years) does it take \$500.00 to double to \$1,000.00 if it is invested in an account paying 6.0% interest compounded quarterly (four times per year)?

    How long (in years) does it take \$500.00 to double to \$1,000.00 if it is invested in an account paying 6.0% interest compounded monthly (twelve times per year)?

    How long (in years) does it take \$500.00 to double to \$1,000.00 if it is invested in an account paying 6.0% interest compounded weekly (fifty-two times per year)?

4. Compare results with your group. Does it appear that the length of the compounding period has an impact on the length of time it takes to double the amount initially invested (if the interest rate and principal remain constant)?

5. Each member of your group should answer one of the questions listed below (you will need a calculator).

    How long (in years) does it take \$500.00 to double to \$1,000.00 if it is invested in an account paying 3.0% interest compounded monthly (twelve times per year)?

    How long (in years) does it take \$500.00 to double to \$1,000.00 if it is invested in an account paying 6.0% interest compounded monthly (twelve times per year)?

    How long (in years) does it take \$500.00 to double to \$1,000.00 if it is invested in an account paying 12.0% interest compounded monthly (twelve times per year)?

6. Compare results with your group. Does it appear that the interest rate has an impact on the length of time it takes to double the amount initially invested (if the length of the compounding period and the principal remain constant)?

7. Which value does your group feel appears to have the biggest impact on the doubling time: the principal, the interest rate, or the compounding period? Why?

Conclusion	Exponential equations can help us to better understand real-life situations. By solving exponential equations created from the compound interest formula it is seen that the doubling time for money invested in an account paying compound interest is affected by both the interest rate and the length of the compounding period.

Name Section Date

Activity 8.7 Fit an exponential function to a set of data.

Focus	Modeling exponential functions
Time	20–30 minutes
Group size	3
Materials	Calculator
Background	In Sections 2.6 and 7.7 of the text, data was modeled by linear functions and quadratic functions, respectively. However, exponential functions are often the best model for situations involving growth. This activity gives you practice finding an exponential function to fit the given data.

1. When discussing stocks that are traded on the New York Stock Exchange, the Dow Jones Industrial Average (DJIA) is frequently used as an indicator of how the stock market is performing. In the mid-1990's, the DJIA experienced what appeared to be exponential growth, as illustrated by the following data.

Year ending	Closing Price
1994	\$3834
1995	\$5117
1996	\$6448
1997	\$7908

(Source: New York Stock Exchange)

2. Use the data points to find a function that fits the data. Since you only need two ordered pairs to find the function, each group member should select the first data point (for the year 1994), and one of the other three data points. Follow the method shown in Example 6 in Section 8.7 of your textbook. Let $t = 0$ represent the year 1994, and write your function in the form

$$P(t) = P_0 e^{kt},$$

where P_0 is the closing price of the DJIA at the end of year 0, $P(t)$ is the closing price of the DJIA in the year t, and k is the exponential growth rate for the DJIA. Show your work in the space provided on the next page. Write neatly so that another group member can follow your steps.

3. Exchange papers with another group member, and check the work on the paper you receive. If you find any errors, discuss them with the person who did the work. He or she should then correct the errors. When you are satisfied with the accuracy of the work, return the paper to the owner.

4. Now, let's determine which function best fits the data. Evaluate your function for the years that were not used to find the function. Show your work below.

   Does your function come close to predicting the closing prices for these years? Compare your prediction with the other group members' predictions. Which function comes closest to predicting the closing prices? Use complete sentences in your answer.

5. If you have the closing price(s) of the DJIA for the years 1998 or beyond, check how accurate your function is in predicting the price(s). As before, compare your prediction with the other group members' predictions. Which function comes closest to predicting the closing prices? Use complete sentences in your answer.

6. Technology Extra: Use a graphing calculator to fit an exponential function to the data. Refer to the Calculator Spotlight of Section 8.7 of your textbook, if needed. Note that all four data points can be used. Then complete the other steps of this activity. Compare this function with the others found using only 2 data points. Which seems best?

Conclusion	As you saw in this activity, data analysis can lead to different answers even though the procedures are correct. This occurs because different sets of data points lead to slightly different quadratic functions. In fact, the more data points you use to find the function, the more accurate the predictions.

Name Section Date

Activity 9.1 Create and simplify equations of circles.

Focus	Equations of circles, completing the square
Time	20 - 25 minutes
Group size	4
Background	Simplifying the equations of circles can be challenging as it frequently involves completing the square (often more than once). Creating and simplifying these equations in a group will give you additional practice with this process and enhance your understanding of these equations.

1. Choose one member of your group to begin the activity. This person should create a circle by deciding where the center will be located and what the radius will be. Write this information down on a piece of notebook paper (for example: Center (-2, 3) and Radius: 3) and pass it to the second group member.

2. The second group member takes the given information and inserts it into the standard form of the equation of a circle given in Section 9.1 of the textbook. Recall that in the standard form of an equation of a circle, $(x-h)^2+(y-k)^2=r^2$, h and k are the x and y coordinates of the center (respectively) and r is the radius. Once the equation is written on the notebook paper (for example: $(x+2)^2+(y-3)^2=9$), pass it on to the third group member.

3. The third group member rewrites this equation on the notebook paper by squaring the binomials and combining like terms. Once the equation is rewritten and simplified (for example, $x^2+4x+y^2-6y+4=0$) transfer it to the Round 1 Equation area below and present it to the fourth group member.

4. The fourth group member uses the method demonstrated in Example 8 of Section 9.1 of the textbook to put the equation back into standard form and identify the center and radius of the circle. This work, and the results, should be written in the Round 1 Solution area below and shown to the first group member to verify the center and radius.

5. Now, repeat the process three more times (Rounds 2 through 4) with a different group member starting the process each time. Be sure that each group member has a turn completing each of the steps. Remember that your equation and solution (with work) should be placed in the appropriate area below when each round is completed.

Round 1 Equation:

Round 1 Solution:

Round 2 Equation:

Round 2 Solution:

Round 3 Equation:

Round 3 Solution:

Round 4 Equation:

Round 4 Solution:

Conclusion	The equation of a circle written in standard form gives the center and radius of the circle. The process of rewriting a given equation of a circle in standard form involves completing the square.

Name Section Date

Activity 9.2 Investigate the effect of varying *a* and *b* on the graph of an ellipse.

Focus	Ellipses
Time	15–20 minutes
Group size	4
Background	The standard equation of an ellipse centered at the origin is $\frac{x^2}{a^2}+\frac{y^2}{b^2}=1$. The graph of the ellipse is an oval-shaped curve, and the exact shape depends on the values of *a* and *b*. In this activity, you will investigate how the shape of an ellipse changes according to the values of *a* and *b*.

1. Each group member should choose one of the following equations. First, determine the values of *a* and *b* for your equation, and write these values in the table. Then, use the techniques given in Example 1 in Section 9.2 of your textbook to graph your ellipse. Use the grid provided on the next page, and label the curve with the equation.

Number	Equation	*a*	*b*
1	$\frac{x^2}{4}+\frac{y^2}{16}=1$		
2	$\frac{x^2}{9}+\frac{y^2}{16}=1$		
3	$\frac{x^2}{16}+\frac{y^2}{9}=1$		
4	$\frac{x^2}{16}+\frac{y^2}{4}=1$		

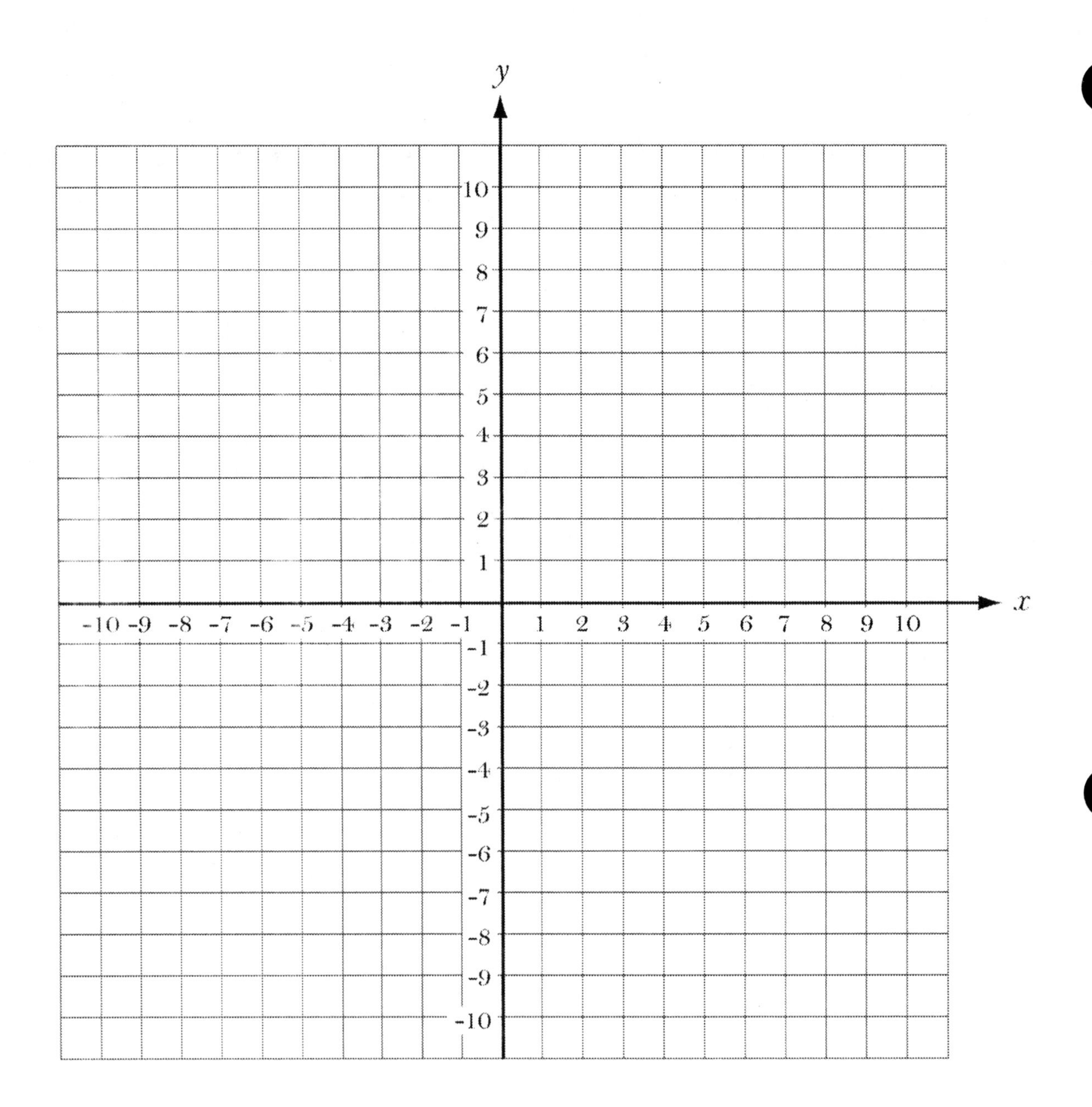

2. When all group members are done, examine the graphs in numerical sequence. What is the effect of the values of a and b on the shape of the graphs? Write your answer using complete sentences.

Conclusion	As you saw in this activity, the graph of an ellipse is elongated horizontally when the value of a is greater than b. When a is less than b, the graph is elongated vertically. You can use this knowledge to help yourself graph ellipses in the future.